# SEPTIC SYSTEMS
# FOR
# CITY SLICKERS

## NANCY MARTIN

Septic Systems for City Slickers

© Nancy Martin

All rights reserved. No part of this publication may be reproduced, stored in a retrieval system, or transmitted in any form or by any means, electronic, mechanical, photocopying, recording or otherwise, without the prior written permission of the author.

National Library of Australia Cataloguing-in-Publication entry

| | |
|---|---|
| Author: | Martin, Nancy, author. |
| Title: | Septic systems for city slickers / Nancy Martin. |
| ISBN: | 9780992367701 (paperback) |
| Subjects: | Sewage--Purification--Australia. |
| | Septic tanks--Australia. |

Dewey Number: 628.3

Published with the assistance of www.loveofbooks.com.au

# INTRODUCTION

This is the beginner's guide to understanding Waste Water Treatment Plants (WWTP), commonly known as septic systems.

Waste water is the general term for all effluent and grey water produced in your home.

It includes flow from:
- Toilet
- Bath
- Shower
- Sinks
- Floor drains
- Laundry
- Dishwasher

The average household uses up to 150 litres of water per person, per day.

In the city, your sewer drains from your house connects to the Council main and flows away.

In non-sewer rural areas, the sewer drain will connect to a WWTP or septic tank.

I will endeavour to explain the different systems, the dos and don'ts of living with a septic system, and a troubleshooting guide for the most common problems and the things you can check to hopefully save you money and some minor headaches.

Let me introduce myself: My name is Nancy and I was a city slicker for 46 years. I never thought about where sewerage went, or the water from the sink or shower, or any of those things while in council main living. Not until something blocked up, and then you rang someone and they fixed it, and I still didn't think about where it went. It just went away. Not my problem anymore. Living on acreage is not that easy.

I met my husband a number of years ago and we have our own business. Our business is called Master Drain Blaster, and specialises in household waste water problems. We pump out tanks, clean drains, locate breaks in pipes, and also patch these breaks where possible. We have trucks that have vacuum units and jet rodding units on the same vehicle. These are a combination unit, and very handy.

John has over 25 years experience in waste management. It didn't take him long to rope me into going in the truck with him and getting knee deep in the business. Actually I enjoy it. We get to work together and I am constantly learning.

Many jobs we go to people say they know nothing about septic systems or how they work, because they too have come from 'council main living', as I call it. John would explain things and having the experience he has he often uses such technical terms, which would even bamboozle me at times. I found myself translating to them and thought I should write a book, so here it is.

I am hoping this will explain things in simple terms. This is only designed to be a simple guide that can hopefully give you the basic understanding of the way these systems work, and perhaps give you a trick or two on getting the problem solved and your system up and running again without the need for costly contractors.

Where does the waste go, you ask? There are Council Water Treatment Plants around where contractors transport only septic waste for recycling.

There are EPA registered recycle plants where all sorts of organic waste can be disposed of. Grease trap and grey water and other waste from paint to egg shells, gets taken to a plant where they mix it with wood chip and screen it many, many, many times, and it eventually becomes potting mix.

# TABLE OF CONTENTS

### Part 1  Septic Systems

1. Bio-systems — 9
2. Septic Tank — 11
3. Sullage Tank – Grey Water — 14
4. Grease Trap — 17
5. Black Water Tank — 19
6. Transpiration (Trench) Box — 20
7. Trenches — 21
8. Do's and Don'ts — 23
9. Troubleshooting Guide — 25
10. Water Tanks — 27
11. Storm Water — 27

### Quick Reference Guide — 28

# Septic Systems

## Bio Systems

These systems usually have one big tank, although some may be more than one. The tank is divided into different-sized chambers. Waste water and solids enter the tank in a single entry point. This is a gravity fed line. This waste in turn moves through the filtered chambers until the final chamber, which is also known as the pump well, uses a pump to expel the treated water, usually out to garden irrigation.

These systems do work well. They also require to be cleaned periodically, but are much more expensive to do than a single septic tank. These systems require to be refilled with water for them to function properly, hence the extra costing.

Quarterly water checks need to be taken and submitted to the council for the PH levels of the water expelled, for obvious health reasons. Most system providers have such treatment plans, which are not overly costly.

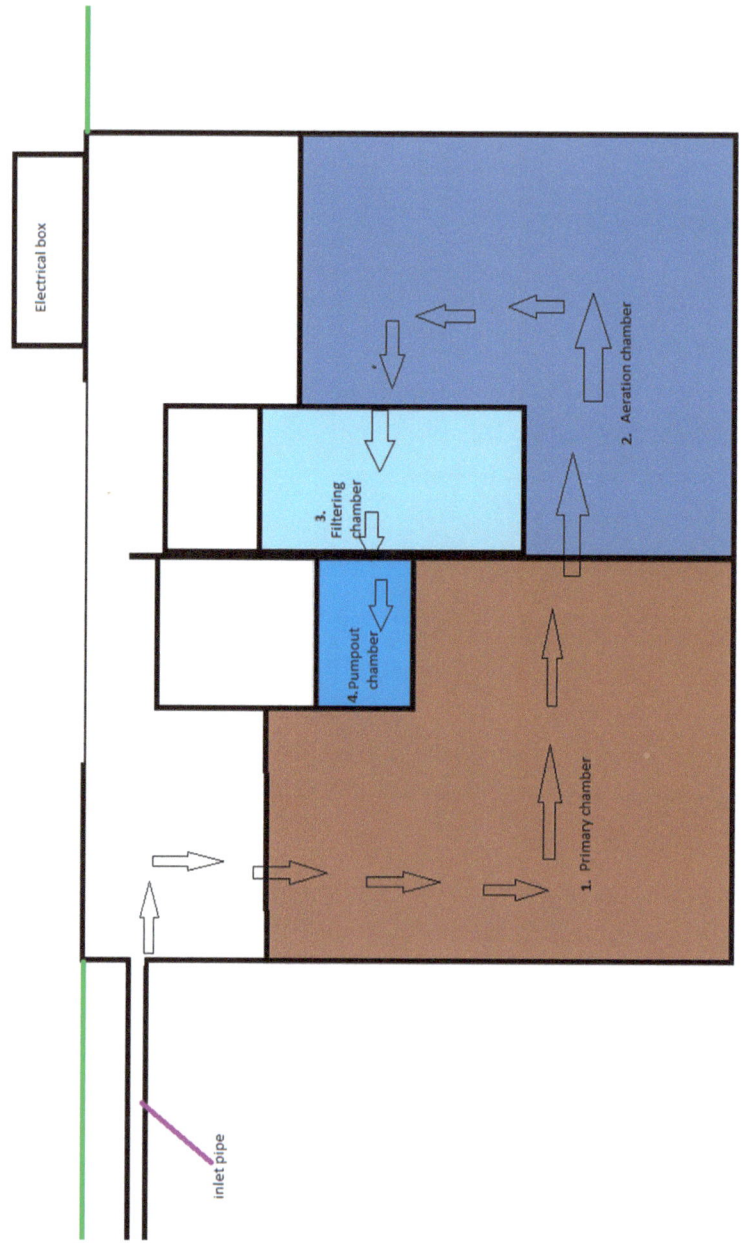

# Septic Tank

A Septic Tank is usually part of a three (3) pit system. Sullage and Grease Trap I will discuss later.

Septic tanks come in all different shapes. The illustrations in this book are for a typical Queensland tank.

A Septic Tank is for toilet waste only. The plumbing from the toilet in the house is gravity fed to the tank. The tank is usually at a lower point than the house level. The tank has an inlet pipe and an outlet pipe. Baffles protect these openings, which also controls the solids entering and exiting the tank. Solids stay on the top of the tank and the liquid stays below. The normal operating level of a septic in good working order is just below the inlet pipe. The liquid leaves the tank via the outlet pipe usually down to a transpiration box, through trenches and seeps out through rubble drains into the ground. Sometimes it just goes straight out to a drain, not via a box.

A Septic system has its own naturally-occurring bacterial growth. When this bacteria level is present, waste is broken down properly and smells and odours are not present.

In this bacterial community, each part eats a different part of the waste. Each level decomposes another until all that is left is the liquid, which leaves the tank via the pipe to the box and trenches. Any interruption to this naturally occurring event can cause an imbalance in the bacterial community. This can stop the waste from being broken down and odour and smells become apparent.

Over time a build-up of sludge will occur in the tank. How often a septic tank needs to be emptied depends upon the usage, the amount of people in the household, the amount of solids, and how well the system is working. A larger family with high traffic and use could require more often pumping than a smaller, low volume place with little traffic. A properly designed and operating system, bar blockages, should be odour free and should last years with minimal maintenance and only periodic inspection.

You can ring contractors, such as Master Drain Blaster, to come and vacuum out the waste and dispose of it for you.

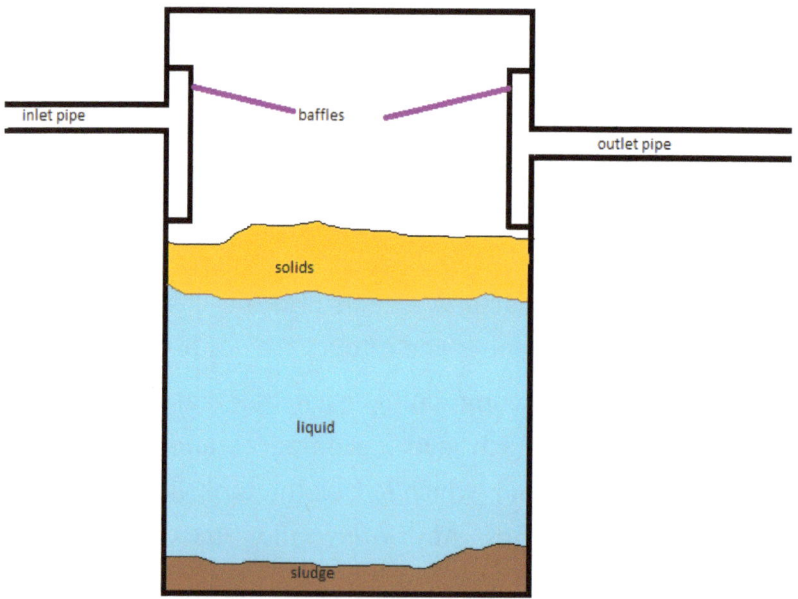

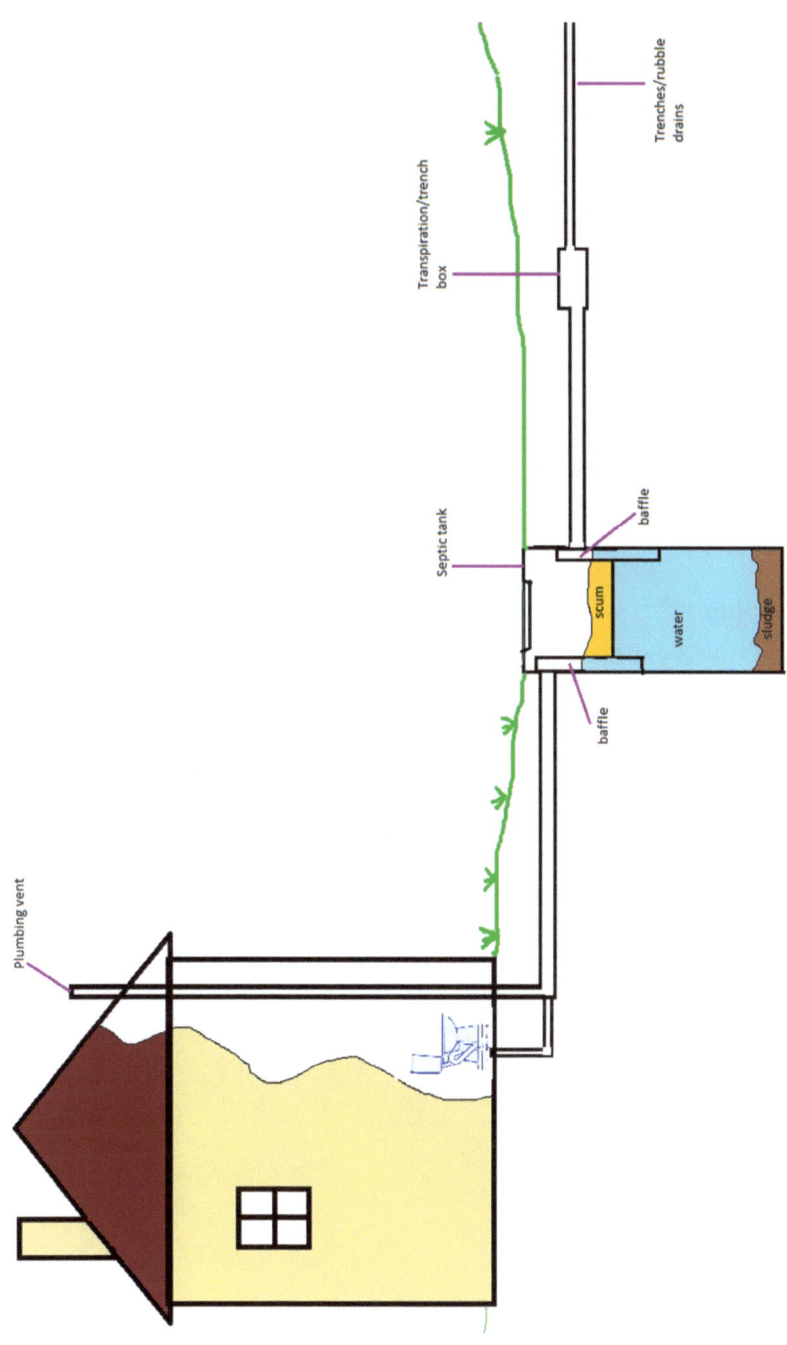

# Sullage Tank – Grey Water

A Sullage tank holds grey water. This is all the left over water from baths, showers, hand basins and washing machines only. No toilet waste enters this tank. In most cases, the grease trap water also goes into the grey water tank. The sullage tank water is pumped out usually to a moveable hose or plumbed irrigation system. This tank will get an amount of sludge build up in the bottom, which should be cleaned out periodically, to prevent the pump from picking up the sludge and doing damage to it. Some systems have a gravity fed line from the sullage out to the ground; this will be downhill from the house, not requiring a pump to expel the water.

You can ring contractors, such as Master Drain Blaster, to come and vacuum out the sludge and dispose of it for you.

This is a picture of a typical grey water pump.

Diagram of piping from kitchen through the grease trap to sullage. Pipe from bathroom and wet areas, through to the sullage tank.

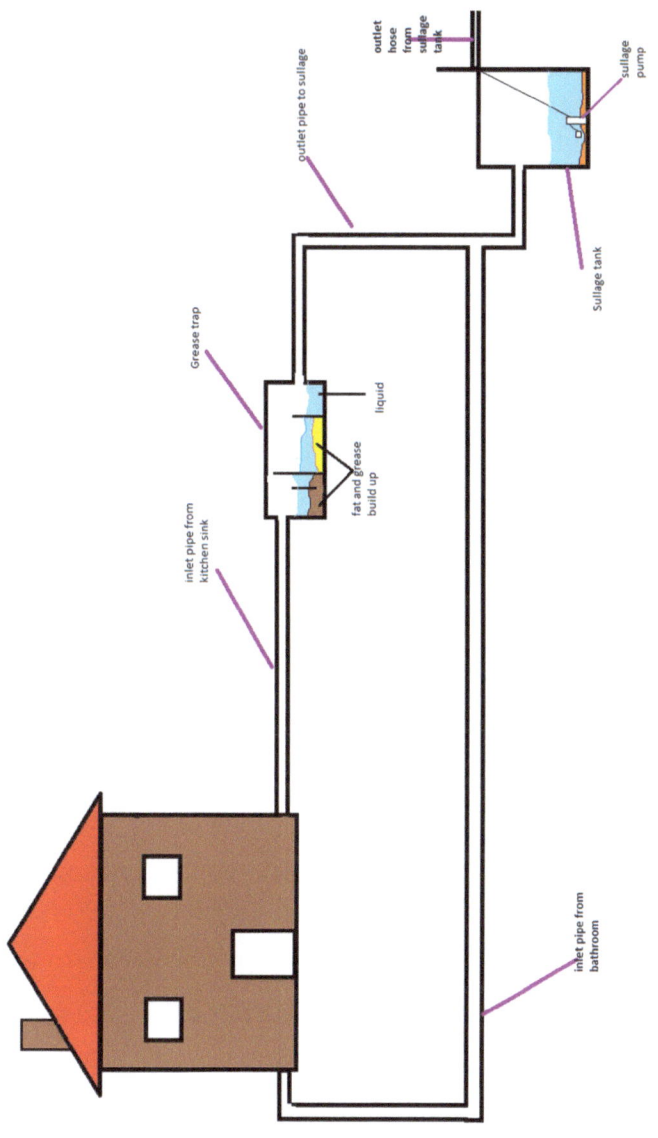

# Grease Trap

The Grease Trap has kitchen sink waste only. The trap is usually located outside the kitchen. In some systems, the tank will be round and sometimes does not go to the sullage; it seeps out to a rubble drain. Most modern traps are rectangle and hold around 140 Litres and have three baffles inside. The inlet side holds the solid fats. Liquid seeps through to centre, where any other solids and grease should stay and the final chamber has an outlet to the sullage tank where leftover liquid flows to. These should be cleaned periodically to maintain a normal level in the trap and sullage tanks. If grease and fat build up in the inlet line, the kitchen sink will run slow. If grease and fats build up in the outlet, this can also block. Fats can also block up the pump if they get across to the sullage tank.

Grease traps can also smell very bad if they build up too much. Food scraps that escape down the sink to the trap, rot, resulting in bad smells. These food scraps produce toxic waste (sulphur gases) – hydrogen sulphide combines with the water present to create Sulphuric Acid. This acid can disintegrate the baffles within the trap and the trap walls causing costly repairs.

You can clean out the grease trap yourself by using a bucket and scraping the solids off the walls or you can call contractors, such as Master Drain Blaster, to do this for you.

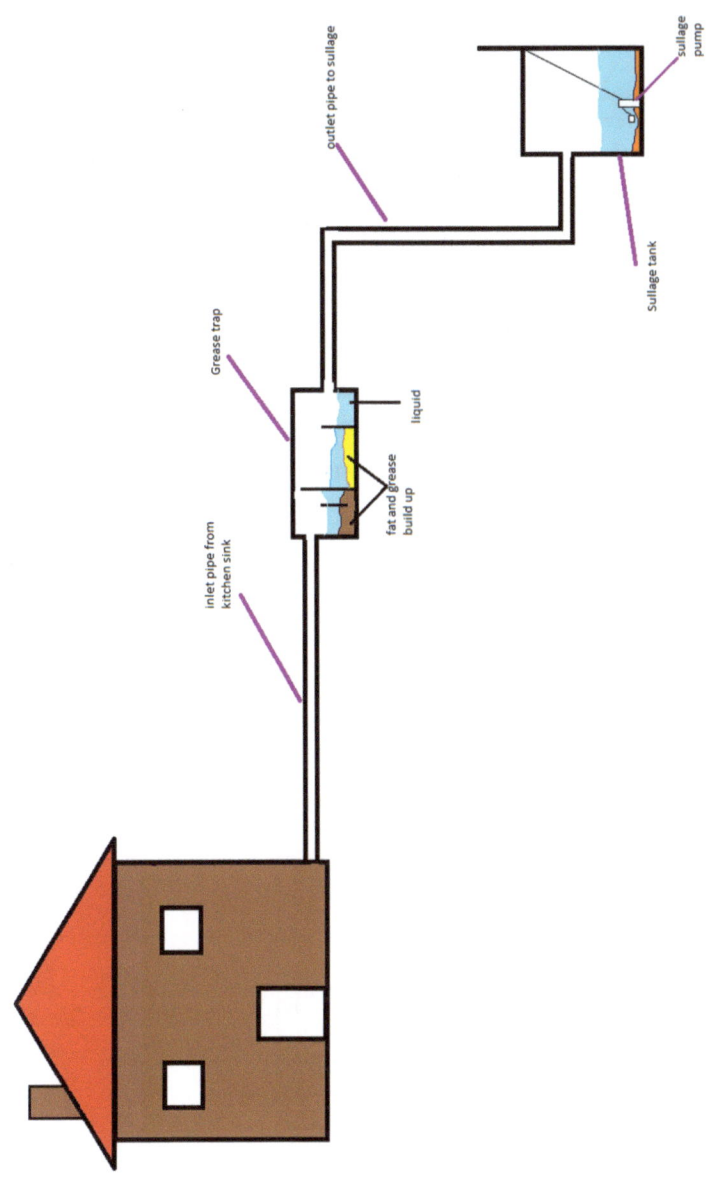

Diagram of pipes from kitchen to the grease trap through to the sullage tank.

# Black Water

Black water is any water containing human faecal waste and urine. It is also known as foul water, or sewage.

Black water tanks are used where gravity feed will not work, if the land is above the house level, the contents need to be pumped uphill. A black water tank will be next to the septic. The septic tank will gravity feed liquid to the black water tank, and then pump from the black water tank up to the transpiration box. These tanks will, over time, need a periodic clean. Like a sullage tank, scum will build up.

You can call contractors, such as Master Drain Blaster, to come and vacuum out the waste and dispose of it for you.

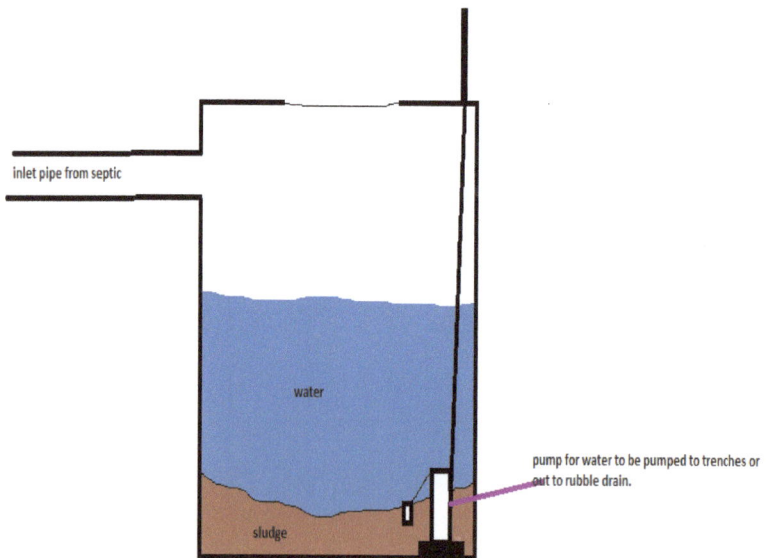

# Transpiration Box

This is also known as a trench box, and this is where the septic drain goes to. Only water from septic tank and black water tanks goes to this. This is the end of the line for septic outlet. This box then goes out to the transpiration trenches. This box will be down the yard from the septic tank as it is usually gravity fed, as described in septic chapter, or pumped in from the black water tank. This box will usually have a 250mm square paver lid on it. If you don't know where this box is, it is usually easily spotted by the very green, grass patches where the trenches go. This box is normally damp when the system is working adequately. If it is dry, then it is usually a blockage in the septic outlet line to the box. If it is full of water, then the trenches are probably not working properly.

# Transpiration Trenches

These are commonly known as trenches, or rubble drains. These are the outlets from the transpiration box. There are nearly always two but sometimes there are three. The water from the septic seeps away down the trenches and soaks into the ground. Unlike sullage water, which can go onto the garden, septic water must go under the ground by law.

The trenches are usually about 600mm deep. From the box there are pipes, which are slotted (Ag Pipe) and encased with small rock/rubble, where the liquid seeps into.

As explained in the last chapter, if the box is full, then the trenches are not working properly. This can be from tree roots, grass roots, building up of calcium, dirt, or foreign blockage. If after being cleaned out the trenches still do not run properly, it is likely the rubble is blocked with septic sludge. In this case, Eco activator* can restore this.

Your trench area should not be driven over, have heavy objects on them, buildings, or even play areas near them. This can damage the trenches and when they have collapsed, they cannot be repaired. Replacing them is a very costly process. They can; however, be cleaned out by high pressure water jet rodding if they are not damaged.

Not all contractors do this, but Master Drain Blaster does specialise in this.

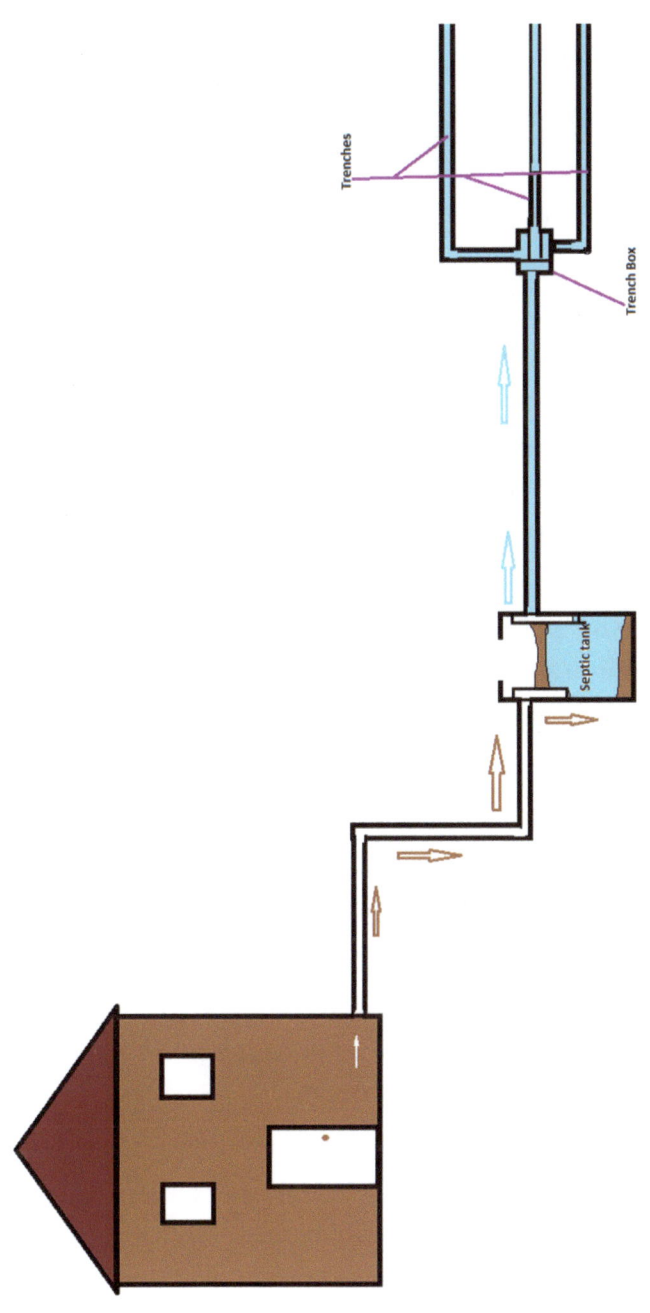

# Do's and Don'ts

No matter what system you are living with, Bio or 3 Pit (Septic, Sullage, and Grease Trap), there are some things that should and should not be done to get the most hassle-free time from your system.

If you are building, it is a good idea not to plant trees or shrubs near the system. These have the potential to creep into the tank as they grow and as the concrete ages and develops small cracks, roots can find their way in and create flow problems due to blocking pipes, inlet, outlet, and trenches as spoken of earlier. Try to clear areas near septics and sullage tanks of trees and shrubs.

Do not flush non-biodegradable items like cigarette butts, sanitary napkins, tampons, nappies, cotton buds, baby wipes, as these items can block up pipes very quickly.

Do not use harsh cleaners with high levels of bleach or caustic; solvents and pesticides are not good for it either. All these will kill the natural bacteria system happening in the tank and smells and problems will occur.

Try and keep the amount of fats and oils to a minimum that goes down the kitchen sink to the grease trap. Regular cleaning of trap is recommended.

Try to keep heavy vehicles and traffic away from the trench area. Continual heavy traffic will collapse the trenches and will result in a costly fix. Try to make yourself familiar with the whereabouts of the trench box and trenches in your yard.

Very high rain fall and flooding can prevent the trenches from working properly, causing flow to back up and not dry out properly.

Use a bio-degradable toilet tissue, on most packs it says whether it is septic friendly and or bio-degradable.

I have used reference to a product called Ecocare Activator. This is an Australian made product which helps to restore and maintain waste treatment systems. Ecocare Activator is a scientifically developed blend of natural ingredients, grease dispersants and biological stimulants that kills odours on contact, cuts through grime, breaks down grease and fat, and restores and maintains a healthy bacterial balance in your system. It is also a fantastic household cleaner so Ecocare Activator provides your home with genuine total treatment protection as it cleans, clears and conditions, all in one go.

# Trouble shooting

| PROBLEM | LIKELY CAUSE | WHAT TO LOOK FOR FIRST | FIX |
|---|---|---|---|
| Blocked toilet | 1. Blockage in inlet pipe to septic<br>2. Blockage in outlet pipe to septic<br>3. Blockage in trenches<br>4. Tank needs pumping out | 1. Check tank level<br>2. Check tank level<br>3. Check trench box<br>4. Check tank level for solids | 1. Jet rod clear pipes<br>2. Jet rod clean pipes<br>3. Get trenches jet rodded and vacuum cleaned<br>4. Pump out tank |
| Slow running kitchen sink | 1. Grease build up in pipe<br>2. Grease trap needs pumping out | 1. Check grease trap level<br>2. Check grease trap level | 1. Jet rod clean pipes<br>2. Pump out grease trap |
| Slow running shower/sink | 1. Hair ball, grease, calcium, soap scum build up in pipes<br>2. Broken sullage pump | 1. Check sullage tank for high level<br>2. Check pump | 1. Get pipes jetted<br>2. Replace sullage pump |
| Smell | 1. Bacteria levels out of sync | 1. Foreign matter introduced into system | 1. Have system pumped out<br>2. Introduce Eco Activator |

| Sullage not emptying | 1. Broken pump 2. Kink in hose 3. Pump not plugged in 4. Float on pump stuck down | 1. Check pump 2. Check hose 3. Check power source to pump 4. Check float | 1. Buy new pump 2. Fix kinks in hose 3. Switch on pump 4. Unhook float |
|---|---|---|---|
| Wet patch in yard | 1. Trench blockage | 1. Check Trench box in yard | 1. Have trench pipes cleaned out 2. Introduce Eco activator to clear out the sludge build up in the rubble |

# Water Tanks and Storm Water

Most acreage properties do not have town water. Therefore, they have to have water tanks to collect rain water for use. Some have what is known as trickle feed, but still require tanks.

Unlike city living, where your gutters and water catchment areas flow through stormwater pipes out to the gutter and down council grates, your water catchments and gutters will be plumbed into your water tank.

These pipes can clog up from time to time; leaves and roots can get into the pipes. Tennis balls, toys, and all sorts of things can also make their way into pipes.

Regular cleaning of gutters and pipes leading to tanks is recommended.

A strainer over pipe ends from gutters, stops a lot of leaves and rubbish entering the pipes. A strainer on the top of your water tank is also recommended to stop rubbish, mozzies and critters entering the tank. A filter is usually fitted between the tank and the house. If you wish, you can then put a filter on the tap in your kitchen sink for your drinking water.

This is a trench box. The jet rodder has pulled back a slug of roots that was blocking the pipe. The vacuum hose sucks out the debris as the cleaning is done.

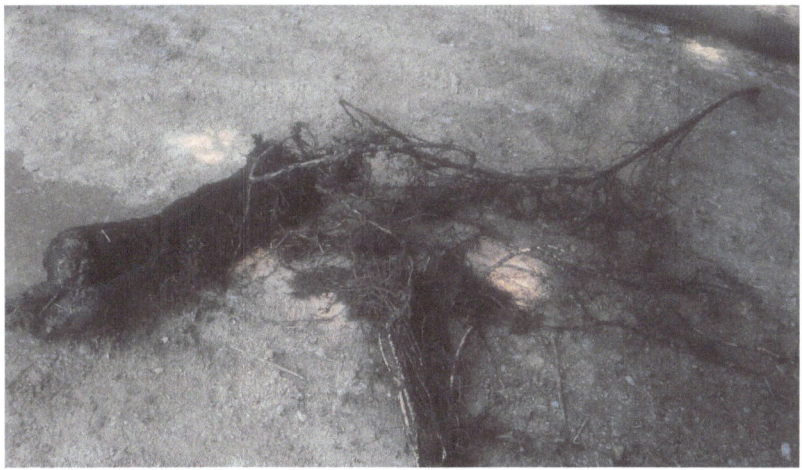

This is a root mat that was pulled out of a storm water pipe.

A septic tank should not look like this. The tank should have mainly liquid with a small layer of solids on top. This tank was solid all the way down, and the level was way too high. As you can see our scraper stood alone in the tank.

More roots out of a storm water drain.

Tennis balls out of a storm water drain. As you can see roots have grown around the balls as well.

Root slug from a storm water pipe.

Storm water pipes being flushed out with the jet rodder.

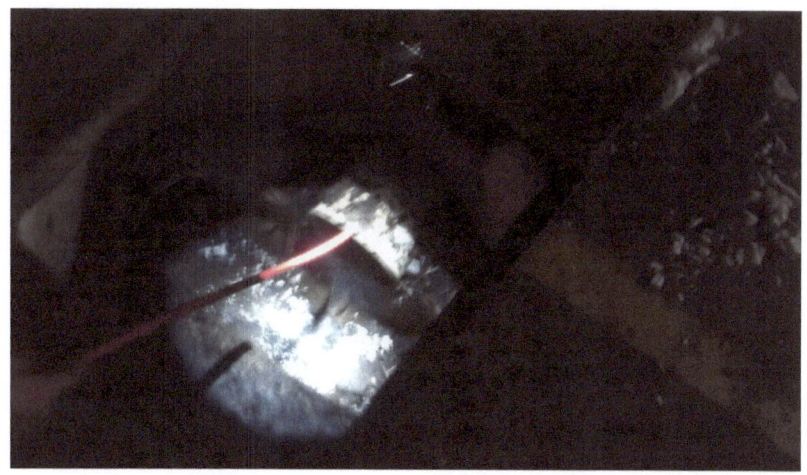

Jetting out a storm water pipe.

This is a classic example of roots growing through and around pipes causing destruction.

This picture is of a crushed pipe. The yellow area to the upper left is a concrete footing. Over time a tree root has grown under the pipe, pushing it up against the footing causing it to crush.

This is one style of Bio System.

Another style of Bio System with filter on top.

Trench box being cleaned out.

Pictures are of a blocked baffle in a septic tank. You can notice the entire choke around the base, this goes up the baffle, blocking the pipe. The baffles in these pictures of a septic tank are way too long. The baffle is touching the bottom of the tank which reduces the flow, therefore making the system back up with solids and it will eventually block up. The baffles were shortened so they would work better.

This is one of our vehicles cleaning out a trench box and trenches.

319 – 327 Lance Road,

North Maclean,

Queensland 4280

| | |
|---|---|
| Phone: | 07 3297 7869 |
| Mobile: | 0432 117 311 |
| Website: | www.masterdrainblaster.com.au |
| Email: | masterdrainblaster@hotmail.com |

www.ingramcontent.com/pod-product-compliance
Lightning Source LLC
Chambersburg PA
CBHW041426190426
43193CB00036B/18